Making Drystone Walls

William Rollinson

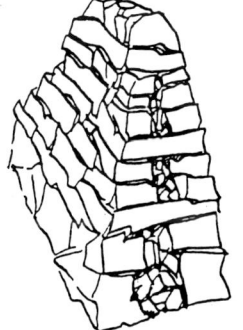

First published in 1998 by
Smith Settle Ltd
Ilkley Road
Otley
West Yorkshire
LS21 3JP

ISBN 1 85825 097 8

British Library Cataloguing-in-Publication data:
A catalogue record for this book is available from the British Library.

Set in Monotype Plantin

Designed, printed and bound by
SMITH SETTLE
Ilkley Road, Otley, West Yorkshire LS21 3JP

Introduction

The wall walks the fell —
Grey millipede on slow
Stone hooves;
Its slack back hollowed
At gulleys and grooves
Or shouldering over
Old boulders
Too big to be rolled away.
Fallen fragments
Of the high crags
Crawl in the walk of the wall.

from *Wall* by Norman Nicholson

Many visitors to the Lake District take away with them an overwhelming impression of a landscape dominated by an irregular pattern of massive drystone walls, as if some cyclopean giant had cast a huge stone net over fell and dale, mountain and moorland. These serpentine stone dikes, which contain no mortar or cement, are an essential element in the landscape, clinging to the fellsides like the rowan and the bilberry, traversing the steepest slopes and surmounting some of the highest summits. They are much a part of the Cumbrian environment as white-washed cottages, packhorse bridges or silent, abandoned slate quarries. So familiar have these walls become that it is difficult to envisage the

Lakeland scene without them; moss-cushioned and lichen-encrusted, they have a mysterious, almost timeless quality, but this is deceptive for, surprisingly, they are a relatively recent addition to the human landscape. In fact, the majority of Lakeland's walls were constructed between 1750 and 1850 when the open, unenclosed moorlands were confined and enmeshed in a network of stone.

It is a common misconception to regard drystone walls simply as field boundaries demarcating individual ownership, for they have other functions. They afford welcome shelter on their lee sides for sheep — and fellwalkers — in wet and windy weather. In some instances they prevent sheep from straying into gullies and becoming 'cragfast' on precipitous slopes — indeed, in the nineteenth century the shepherds of Ennerdale kept in repair a wall which prevented sheep from wandering onto the front of Pillar Rock, and a similar wall along the summit of Dow Crag served the same function. Elsewhere, drystone walls form *outgangs,* wide-mouthed funnels narrowing down to parallel walls leading to sheep pens. On the floors of several of the larger valleys such as Great Langdale, Wasdale and Borrowdale, the walls not only divide the land into a jigsaw of irregular fields, they also act as stone-dumps, for this 'inland' pasture has been created by the back-breaking process of clearing the area of boulders, and simplest way of disposal was to build them into bastion-like walls many feet thick.

To the layman's eye, drystone walls are simply uncemented stones balanced precariously on top of each other — but nothing could be further from the truth. They are basically structures in equilibrium in which the load is transferred downward through each carefully-laid course onto large *footing stones* at the base of the wall, cohesion being achieved by placing one stone on two and two stones on one — *'yan on twa an' twa on yan'* was the waller's guiding maxim. Sharp-eyed fell-wanderers will have appreciated that most drystone walls are in effect two walls with a gap in the centre filled with *hearting stones,* the two faces being tied together 'like Cumberland and Westmorland champion wrestlers' (Norman Nicholson) by *through stones* projecting out from the wall on both sides. As the wall increases in height, so the width decreases from about 3 feet (90cm) to 1-2 feet (30-60cm) at the top, the 'batter' being achieved by setting the stones into the wall as the construction proceeds. Finally, the top of the wall is finished by a line of *cam stones* forming a rough, angular outline which discourages the unwelcome attention of *lish* (agile) mountain sheep as well as the clumsy clambering of hoards of walkers.

It follows, of course, that the wallers used stone found in the immediate vicinity of the wall, and this was often obtained from small quarries above the point at which they were working so that the stone could be dragged *down* the fellside. It also follows that Lakeland's walls closely reflected the great variety of solid geology, from the craggy, angular walls of the volcanic rocks of the central fells, the more fissile slate walls of the country around Coniston Water and Windermere, the pink, crystalline granite walls of Eskdale, the silver-grey limestone walls which ring the central uplands, and the turf-and-cobble walls of the Cumbrian coast. Armed with a geological map, the fellwalker should be able to identify the major Lakeland rock types simply by observing the drystone walls ...

Without documentary or cartographic evidence, the dating of walls proves difficult. The turf-and-stone wall in upper Eskdale, built by the Cistercian monks of Furness Abbey to enclose their sheep pasture, can be dated by monastic documents to the period between 1284 and 1290. Similarly the wall which snakes across Red Screes, dividing the townships of Ambleside and Troutbeck, is documented to 1551. Elsewhere, enclosure awards and maps allow accurate dating where such documents survive and, in some instances, dated gate-stoups pinpoint the year a wall was built, but these are exceptions.

Sadly, little is known of the individual wallers who transformed the fellsides during the eighteenth and nineteenth centuries. Certainly their working lives were arduous and harsh; frequently they bivouacked on the fells close to their work, walling from sun-up to sunset, and coming down to the valley settlements only at weekends. Wages were low; in 1774 a rood, or seven yards (6.5m), of wall five and a half feet (1.5m) high could be built for between 1s 6d and 1s 8d per rood, but by 1877 this had increased to 6s 6d per rood. Many wallers were illiterate and could sign their meagre wage receipts merely with a cross — but, in other ways, these nameless craftsmen stamped their signature on the Lake District landscape for everyone to see and admire today.

Acknowledgements

Friends of the Lake District for the photo on page 45, David E Kirk for the drawing on page 9 and Peter Kearney for the inset on page 31. The poem on page 3 and the quotation on page 23 are copyright the literary executors of Norman Nicholson.

The author would like to thank David Kirk, Irvine Hunt, Peter Kearney and Christine Denmead for their help in compiling this booklet.

Wasdale Head enmeshed in a network of drystone walls. The older, irregular field boundaries on the flat valley floor contrast with the more regular late eighteenth and early nineteenth century *intake walls* on the fellside.

An eagle's eye view of the contrasting patterns of drystone walls at Stool End, Great Langdale.

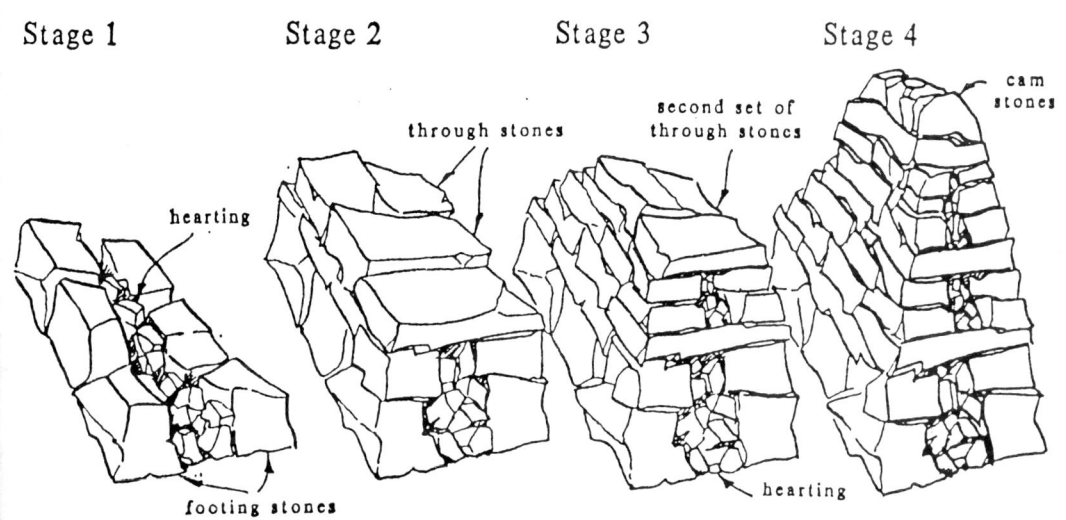

Stage 1   Stage 2   Stage 3   Stage 4

hearting

through stones

second set of
through stones

cam stones

footing stones

hearting

Stages in the building of a drystone wall. Note that the wall is thicker at the base than at the top.

Cross-sections through drystone walls:
the through stones, which project through the
wall and tie the two outer faces together, are
conspicuous, though in the example on the
right the hearting stones in the centre of the
wall have fallen away.

The low turf and stone wall in upper Eskdale, on which the walker is standing, was built as part of a sheep enclosure by the Cistercian monks of Furness Abbey in the late thirteenth century.

The wall which snakes over the fell below Red Screes on Kirkstone Pass divides the Township of Ambleside (left) from the Township of Troutbeck (right). The wall can be dated to 1551.

The angular blocks which form this wall on Low Pike, north of Ambleside, are typical of the resistant rock which makes up the rugged central fells.

Ingenuity in wall building: an exposure of hexagonal columnar volcanic rock in the upper Lickle Valley has provided the waller with the means to bridge a small beck near Stephenson Ground Farm.

The platey, fissile slates of the southern fells have been used for this wall on the Garburn Road near Troutbeck, Windermere. The marked tendency of the rock to split into thin slabs allows water to penetrate into the wall, making this an unsatisfactory walling material.

In areas where flagstones outcrop, fences made of interlocking flags may be found. This example is in the Lickle Valley, but there are others at Coniston, Hawkshead and Ambleside.

A granite wall in Eskdale. The boulders have been partially rounded either by river action or glaciation.

Around the Cumbrian coast and on Walney Island, water-worn *cobbles* (boulders) provide the building material for many walls. This example, near Bootle, is made up of alternate layers of turf and stone.

A sandstone wall, west Cumbria.

A 'cyclopean' wall near Ulpha, Dunnerdale. These enormous boulders are probably the result of field clearance. The folded Ordnance Survey map at the centre of the photograph gives an indication of the scale.

This unusual wall near Burnmoor Tarn, between Wasdale and Eskdale, defies all the accepted rules of drystone wall building. Common in the west of Ireland, walls like these reduce the effects of wind pressure, but fellwalkers beware — they cannot be scaled without dire consequences!

Bastion-like walls at Wasdale Head. The walls here are not only field boundaries but also depositories for thousands of tons of boulders cleared from the surrounding pastures by men, women and children.

'*... shouldering over old boulders too big to be rolled away*' (Norman Nicholson). In effect this boulder in Mosedale has saved the waller both time and effort, and as such would have been welcomed.

To prevent the wall slipping down the fellside, the long axes of the stones remain horizontal despite the steep slope of the land.

A *hogg hole* near Coniston. In order to allow *hoggs* or *hoggets* (yearling sheep) free access from one pasture to another, holes were left in the base of drystone walls. These could be blocked if necessary by a slate flag or a boulder.

Two unusual hogg holes in the Lickle Valley. In order to build the one above, a temporary wooden framework has been constructed and the stones chocked around it. The hole on the left has been blocked to prevent use.

ee *boles* at Hodge Close near Tilberthwaite. Square and rectangular holes in the drystone wall
housed straw bee *skeps*. Most bee boles face south to make maximum use of the available
sunlight and also to afford shelter from the cold north wind.

One of a series of gun holes used for shooting activities, at Seathwaite, Dunnerdale.

Right: a *wall head*. The visible line demarcates ownership when a common wall divides land owned by two parties. If the wall *rushes* (collapses), responsibility for repair can be determined

Here a wall in Grizedale
Forest has collapsed
exactly on the wall head.

A drystone fox trap above Levers Water, near Coniston.
The trap was baited with a tasty morsel dangled from
the end of a plank like the clapper in a bell; the fox
'walked the plank' and was precipitated into a stone
bell-shaped prison from which there was no escape.

Frost damage to a limestone wall.

Here quarry offcuts have been used to make an unsightly repair to a wall near Rydal.

This wall at Nether Wasdale has had the cam stones cemented on top. When the wall rushed (collapsed), the cams were left suspended like beads on a string.

Defying gravity, this
drystone wall bridges
a small beck near Boot
in Eskdale. There is no
lintel and the stones
have been chocked
together.

Dated gate stoups. This 1766 stoup is near the church at Ulpha in Dunnerdale. The initials are those of John Gunson.

An early stoup at Carter Ground in the Lickle Valley. The name 'Carter' and the date 1663 can be deciphered.

An elaborately carved and dated stoup near Taw House in Eskdale. The date is 1817 and the initials are those of John Vicars Towers.

Right: the date 1798 and the initials J B appear this example from Rosthwaite in Borrowd

The destruction of a drystone wall. The building of the Kendal bypass resulted in the removal of miles of drystone walls.

This once silver-grey limestone wall near Arnside has been sprayed with tar and diesel oil to discourage visitors from removing the limestone for use in garden rockeries.

A lesson in
ergonomics: [a]
waller will alwa[ys]
maximise his ef[fort]
by carefully
selecting his sto[ne]
before finding [a]
place for them [in]
his wall. Here t[he]
waller is infilli[ng]
with small hear[ting]
stones.

Johnny Birkett of High Yewdale Farm near Coniston *gaping* (repairing) a length of drystone wall.

Chris Kirkby and Jimmy Spedding rebuilding a wall at Wasdale. Normally wallers worked in two-men gangs, one man on one side of the wall and one on the other.

n annual drystone walling competition is organised by the Friends of the Lake District in an attempt to encourage this ancient craft.

The next generation: youngsters try their hand at drystone walling at Brockhole Visitor Cent
Windermere, learning the lesson that it's not as easy as it first appears.

October 12 To 3 days Work for Walking Sheap fauld wall — — ..10..6
16 To 3 days Work for Getting dore soales & Paging stones — . 11–10–6
22 To 6 days Work for Getting haland stones — — — –1–1–0
25 To 3 days Work for Laading haland stones & Gate Post 11–10–6
29 To 4 days Work for Setting Do, & dore soales — 11–14–0
Novmb 2 To 4 days Work for walling Payper wall & Plastring Jither 11–14–0
26 To 2 days Work for setting Gate Post & Plastring June House– 7–0

£ 16–2–0

Dec 24th 1841 Settled
Jos Lancaster his
mark

seph Lancaster was a nineteenth century waller whose work may be seen in the Lanthwaite a of Crummock. He was illiterate and could only sign his wage receipt with a simple cross — but in another way he has indelibly stamped his signature on the fells.